Rupture of a Railroad Tank Car Containing Hazardous Waste
Freeport, Texas
September 13, 2002

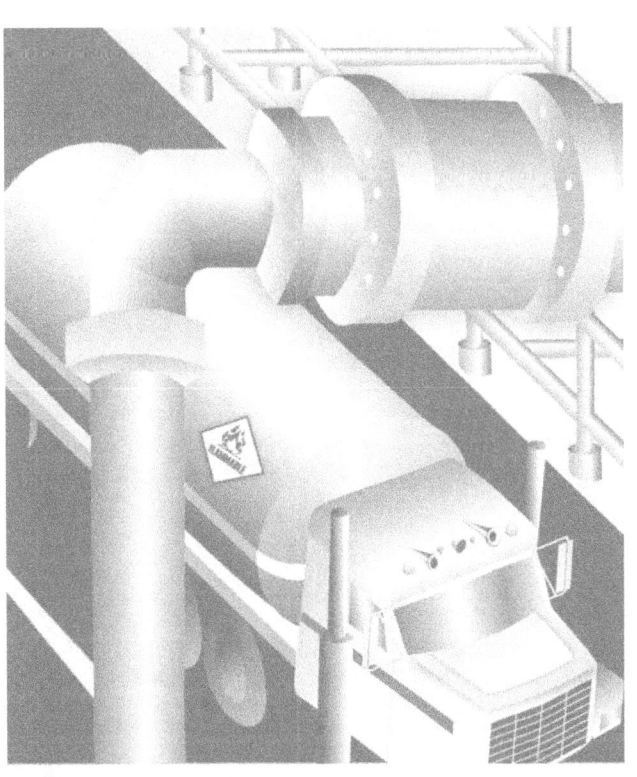

**Hazardous Materials
Accident Report**

NTSB/HZM-04/02

PB2004-917003
Notation 7661

**National
Transportation
Safety Board**

Washington, D.C.

this page intentionally left blank

Hazardous Materials
Accident Report

Rupture of a Railroad Tank Car Containing Hazardous Waste
Freeport, Texas
September 13, 2002

NTSB/HZM-04/02
PB2004-917003
Notation 7661
Adopted December 1, 2004

National Transportation Safety Board
490 L'Enfant Plaza, S.W.
Washington, D.C. 20594

National Transportation Safety Board. 2004. *Rupture of a Railroad Tank Car Containing Hazardous Waste, Freeport, Texas, September 13, 2002.* **Hazardous Materials Accident Report NTSB/HZM-04/ 02. Washington, DC.**

Abstract: About 9:30 a.m. central daylight time on September 13, 2002, a 24,000-gallon-capacity railroad tank car, DBCX 9804, containing about 6,500 gallons of hazardous waste, catastrophically ruptured at a transfer station at the BASF Corporation chemical facility in Freeport, Texas. The tank car had been steam-heated to permit the transfer of the waste to a highway cargo tank for subsequent disposal. The waste was a combination of cyclohexanone oxime, water, and cyclohexanone. As a result of the accident, 28 people received minor injuries, and residents living within 1 mile of the accident site had to shelter in place for 5 1/2 hours. The tank car, highway cargo tank, and transfer station were destroyed. Two storage tanks near the transfer station were damaged; they released about 660 gallons of the hazardous material oleum.

The National Transportation Safety Board identified one major safety issue, the adequacy of procedures for heating hazardous materials cargoes in railroad tank cars before transfer.

As a result of its investigation of this accident, the Safety Board makes safety recommendations to the Research and Special Programs Administration, the Occupational Safety and Health Administration, and the Environmental Protection Agency.

Contents

Executive Summary

About 9:30 a.m. central daylight time on September 13, 2002, a 24,000-gallon-capacity railroad tank car, DBCX 9804, containing about 6,500 gallons of hazardous waste, catastrophically ruptured at a transfer station at the BASF Corporation chemical facility in Freeport, Texas. The tank car had been steam-heated to permit the transfer of the waste to a highway cargo tank for subsequent disposal. The waste was a combination of cyclohexanone oxime, water, and cyclohexanone. As a result of the accident, 28 people received minor injuries, and residents living within 1 mile of the accident site had to shelter in place for 5 1/2 hours. The tank car, highway cargo tank, and transfer station were destroyed. The force of the explosion propelled a 300-pound tank car dome housing about 1/3 mile away from the tank car. Two storage tanks near the transfer station were damaged; they released about 660 gallons of the hazardous material oleum (fuming sulfuric acid and sulfur trioxide).

The National Transportation Safety Board determines that the probable cause of the rupture of railroad tank car DBCX 9804 was overpressurization resulting from a runaway exothermic decomposition reaction initiated by excessive heating of a hazardous waste material. Contributing to the accident was the BASF Corporation's failure to monitor the temperature and pressure inside the tank car during the heating of the hazardous waste.

The Safety Board identified the following safety issue during this investigation:

- Adequacy of procedures for heating hazardous materials cargoes in railroad tank cars before transfer.

As a result of its investigation, the Safety Board makes safety recommendations to the Research and Special Programs Administration, the Occupational Safety and Health Administration, and the Environmental Protection Agency.

Factual Information

The Accident

Synopsis

About 9:30 a.m. central daylight time[1] on September 13, 2002, a 24,000-gallon-capacity railroad tank car, DBCX 9804, containing about 52,450 pounds (6,500 gallons)[2] of hazardous waste, catastrophically ruptured at the BASF Corporation (BASF) chemical facility in Freeport, Texas. The tank car had been undergoing steam-heating to permit the transfer of the waste to a highway cargo tank for subsequent disposal. The waste was a combination of cyclohexanone oxime, water, and cyclohexanone. The railroad tank car, the highway cargo tank, and the transfer station were destroyed. (See figure 1 for a postaccident photo of the transfer area.) Two nearby storage tanks containing oleum (fuming sulfuric acid and sulfur trioxide) were damaged and released about 10,650 pounds (660 gallons)[3] of material. Twenty-eight people received minor injuries, and residents living within 1 mile of the accident site had to shelter in place for 5 1/2 hours.

Events Preceding the Accident[4]

Generation and Initial Storage of the Waste in Freeport. On June 1, 2001, a process upset[5] occurred at the caprolactam II[6] process of the BASF facility in Freeport. On June 5, 2001, as a result of the upset, about 90,875 pounds (11,261 gallons) of material, consisting of about 94 percent cyclohexanone oxime, 4 percent water,[7] and 2 percent cyclohexanone, was temporarily transferred to four highway cargo tanks. On June 11, 2001, the material was to be transferred to railroad tank car DBCX 9804, but it had solidified within the highway cargo tanks. (The material would solidify at temperatures below 194° F.) BASF applied steam heat from a facility boiler to the four highway cargo tanks to liquefy the material for transfer. On June 12, 2001, BASF transferred the contents of the four highway cargo tanks to tank car DBCX 9804.

---

[1] Unless otherwise specified, the times used in this report are central daylight time.

[2] According to the BASF Corporation, the density of the waste material was about 8.07 pounds per gallon.

[3] The density of oleum is about 16.14 pounds per gallon.

[4] See appendix B for a summary chronology of significant events preceding the accident.

[5] A *process* is a facility area where hazardous materials are used, stored, manufactured, or handled. A *process upset* is a condition outside of routine operating parameters in chemical manufacturing.

[6] Caprolactam is used in the production of nylon. There are two caprolactam processes at the Freeport facility, known as caprolactam I and caprolactam II.

[7] Because cyclohexanone oxime is produced in water within the process, some water accompanies it.

Figure 1. Damage to transfer area caused by catastrophic rupture of tank car DBCX 9804.

Sometime between June 12, 2001, and April 26, 2002, BASF applied steam heat to the tank car to liquefy the waste mixture so water could be removed and disposed of on-site.[8] BASF was unable to provide information on how many times the tank car was heated, the specific dates of the heating, the pressures at which the steam heat was applied, or the amount of water removed. According to BASF, no additional materials were added to the tank car during this period.

On February 12, 2002, a BASF contractor took samples of the material in DBCX 9804. The samples were sent to several waste disposal companies, which provided BASF with disposal cost estimates. BASF subsequently contracted with Missouri Fuel Recycling Environmental Services (MFR) in Hannibal, Missouri, to dispose of the hazardous waste. (MFR conducted tests on its sample, which determined the material's heating value [13,600 British thermal units per pound] and the fact that it contained no metals.)

On or about April 2, 2002 (exact date unknown), BASF removed about 90,875 pounds (11,261 gallons) of waste material, consisting of about 94 percent cyclohexanone oxime, 4 percent water, and 2 percent cyclohexanone, from a second

---

[8] Environmental regulations allow facilities to dispose of some types of waste materials, such as the water from DBCX 9804, on-site.

process upset of the caprolactam II process and loaded it into four highway cargo tanks. According to BASF, the second batch of material had essentially the same chemical composition as the material from the June 1, 2001, process upset.

On April 26, April 29, and May 10, 2002, BASF transferred the second batch of material from the highway cargo tanks to DBCX 9804, which still contained the material from the June 1, 2001, caprolactam II process upset. According to BASF, after the second batch of material was added to DBCX 9804, nothing more was added to or removed from the tank car while it remained at the Freeport facility.

On June 21, 2002, BASF released railroad tank car DBCX 9804, containing the material from both process upsets, to the Union Pacific Railroad for transportation to MFR in Hannibal, where the waste was to be unloaded and incinerated. According to BASF, when the tank car departed Freeport, it contained about 181,750 pounds (22,522 gallons) of the waste material.

Heating and Unloading the Waste in Hannibal. On the morning of July 7, 2002, tank car DBCX 9804 arrived at the Continental Cement facility in Hannibal. MFR, a division of the Continental Cement Company, procures hazardous wastes as supplemental fuels for the cement production process at the facility. Each year, MFR unloads about 350 railroad tank car loads of such waste at Continental's Hannibal plant. MFR had never before heated or unloaded the waste material contained in DBCX 9804.

On July 8, MFR began preparing DBCX 9804 for unloading by moving the tank car to a boiler station, where the waste material within the tank car was to be heated to liquefy it. An MFR liquid fuel supervisor who oversaw most of the unloading stated that when the tank car arrived at the boiler station, he looked through a 6-inch-diameter access opening in one of the two dome housings on the top of the tank car. The liquid fuel supervisor said he saw a white, solid material in the tank car. He attempted to take a sample using a 7-foot-long sampling tube but was unable to do so because the material was too hard. He said the material's hardness was comparable to steel.

The liquid fuel supervisor decided to try to take a sample from a different location. After inserting the sampling tube through a 3-inch-diameter eduction (liquid or product) line valve, the liquid fuel supervisor extracted a sample through the eduction line, which extended almost to the bottom of the tank car. The liquid fuel supervisor described the sample material extracted as "flaky, gooey, very thick, black, and the consistency of oatmeal."

On the morning of July 9, MFR began to steam-heat DBCX 9804 using a natural gas boiler to generate steam. The steam was applied to the tank car at a pressure of 115 pounds per square inch, gauge (psig). MFR inserted a pressure/temperature gauge (attached to a 4-foot probe) into the 3-inch valve opening on top of the tank car. The gauge was connected to an automatic shutdown mechanism on the boiler that would stop the heating if the pressure within the tank car exceeded 15 psig.

MFR employees documented heating, unloading, and other daily activities in an operations logbook. According to the logbook, at 9:00 a.m. on July 9, the temperature of the material in the tank car was 107° F. At 2:00 p.m., the temperature was 119° F and the pressure within the tank car was 3.5 psig. The heating continued until the boiler was shut off at 9:30 p.m. on July 9.[9] According to the logbook, when the boiler was shut off, the temperature of the material was 131° F and the pressure within the tank car was 0.3 psig.

At 5:00 a.m. on July 10, MFR restarted the boiler and resumed the heating process. According to logbook notes, at 9:00 a.m., the temperature of the material was 146° F, and the material was "still chunky." At 12:00 p.m., the temperature of the material was 150° F. At 1:00 p.m., the liquid fuel supervisor, finding that the 4-foot probe attached to the gauging device did not extend to the bottom of the tank car and so was not providing an accurate reading of the material's overall temperature, temporarily removed the probe and replaced it with a 10-foot metal rod.

The liquid fuel supervisor stated that, after leaving the rod in place for about 5 minutes, the rod's temperature was 129° F (21° F lower than the most recent reading obtained using the 4-foot probe attached to the gauging device). When the liquid fuel supervisor removed the rod, he could see the material was less viscous, but he also noted that the material hardened very quickly when exposed to the ambient temperature. The pressure/temperature gauge was reinserted into the tank car, and the heating continued until 11:00 p.m. on July 10, when the heating was discontinued for the night. At this time, the temperature of the material was 195° F.

At 5:00 a.m. on July 11, the temperature of the material in the tank car was 215° F.[10] The liquid fuel supervisor visually checked the material's viscosity and determined that the material could be unloaded. MFR moved the tank car to a transfer station about 200 yards from the boiler station. When the tank car arrived at the transfer station, MFR employees used a vacuum truck to prime the pump to start the flow of product from the tank car to a highway cargo tank. The transfer continued until the highway cargo tank was full, at which time it was weighed. The highway cargo tank contained about 57,840 pounds (7,167 gallons) of waste material.

The highway cargo tank was moved to another area where its contents were unloaded into a storage tank; then it was returned to the transfer station, and the transfer of material from the tank car resumed. The liquid fuel supervisor stated that while material was being unloaded from the tank car to the highway cargo tank for the second time, the pump began to "suck air," which indicated to him that all the material that could be pumped from the tank car had been removed. The vacuum truck was used to suck out any additional material in the transfer hoses. Then, the pump was turned off, the hoses were disconnected, and the transfer process was terminated.

The highway cargo tank and the vacuum truck were driven to the storage tank area, where they were weighed, and the waste was transferred to the storage tank. The second

---

[9] MFR did not operate a third shift at the Hannibal facility, so the boiler was shut down overnight, when no employees were on duty.

[10] In an insulated tank car, the temperature of heated material can continue to rise even after heating has been discontinued.

load in the highway cargo tank contained about 36,540 pounds (4,528 gallons) of waste, and the vacuum truck contained about 6,220 pounds (771 gallons) of waste. According to the load receipt, MFR unloaded 100,600 pounds (12,466 gallons) of material from DBCX 9804. The unloading procedure was completed about 4:00 p.m. on July 11. According to the secondary liquid fuel supervisor, shortly after 4:00 p.m., he closed all the valves and access ways on DBCX 9804 and applied new security seals to the tank car's two dome housings.[11] DBCX 9804 left Hannibal on July 12. While traveling from Hannibal to Freeport, DBCX 9804 encountered no unusual delays or routing.

When BASF sent tank car DBCX 9804 to MFR for unloading, it had provided MFR a State of Missouri hazardous waste manifest for the shipment. The manifest from BASF did not record the total quantity of material sent in the tank car. When MFR returned the manifest to BASF, it was signed by an MFR secondary liquid fuel supervisor and dated July 11, 2002. The manifest returned from MFR stated that 91,013 pounds (11,278 gallons) of material had been in the tank car when MFR received it. Neither the primary nor secondary liquid fuel supervisor, nor any other MFR employee, could tell investigators how MFR had derived the figure for total quantity received from BASF that appeared on the manifest. The manifest also stated that MFR had unloaded 90,973 pounds (11,273 gallons) of material at Hannibal.

Heating and Unloading the Waste in Freeport. Tank car DBCX 9804 arrived at the BASF Freeport facility on July 22. Facility personnel conducted a routine inspection of the tank car and found no security seals on the dome housings or anywhere else on the tank car. Aside from the missing seals, they found no indications of tampering with the tank car or its contents. Also as part of the inspection, they removed the 6-inch access plate and examined the interior of the tank car. A BASF contract employee who looked into the tank car observed that it was still about one-third full of material. BASF accepted the tank car. BASF did not contact MFR concerning the material remaining in the tank car or the absence of security seals.

Facility personnel determined that the tank car could not be cleaned (in the typical manner) because it contained too much material. They decided to transfer the material remaining in DBCX 9804 to highway cargo tanks for disposal. Between July 22 and September 11, no action was taken to unload the tank car. At some time during this interval, a BASF employee and a Mundy Industrial Services (Mundy)[12] employee looked into the tank car through the 6-inch access opening and found that the tank car was about one-third full of material. Both employees stated that they saw liquid in the tank.

On September 11, DBCX 9804 was moved to the caprolactam II process area so it could be prepared for heating and unloading. A 7,000-gallon-capacity highway cargo tank was brought to serve as a receiving vehicle for the waste. The highway cargo tank was already about half full of a similar waste material.

---

[11] There is no Federal requirement that security seals be attached; generally, companies establish their own practices with respect to security seals. MFR practice was to attach security seals to all outgoing tank cars, even those MFR believed were empty.

[12] The contractor Mundy Industrial Services of Houston, Texas, provides chemical loaders, a track repair crew, clerical support for truck shipments, and security inspections for BASF.

In the afternoon of September 11, two Mundy employees—a primary unloader and a secondary unloader—connected the heating and unloading hoses to the railroad tank car. They threaded one end of a 3-inch-diameter hose into the 3-inch eduction line valve on top of the tank car; they inserted the other (open) end of the hose into the manway opening on the top of the highway cargo tank. They connected a second hose from a facility nitrogen supply to a 2-inch-diameter vapor line valve[13] on top of the tank car. The nitrogen was available to provide additional pressure to the tank car, if needed, to assist in unloading the waste. The nitrogen was not turned on; the eduction and vapor line valves on the tank car were not opened.

Unloaders ran a steam hose between a facility boiler and the inlet for the tank car's heating coil. They placed no pressure or temperature gauges in the tank car to monitor the interior conditions, nor did they employ any automatic shutdown mechanism. Later in the afternoon of September 11, steam was applied to the tank car at 60 psig. After about 1 hour of heating, facility personnel decided to adjust the car's position. They disconnected the unloading and heating lines, repositioned the tank car, and reconnected the unloading and heating lines. They did not resume heating the tank that day. (See figure 2 for a diagram showing the layout of the Freeport transfer area.)

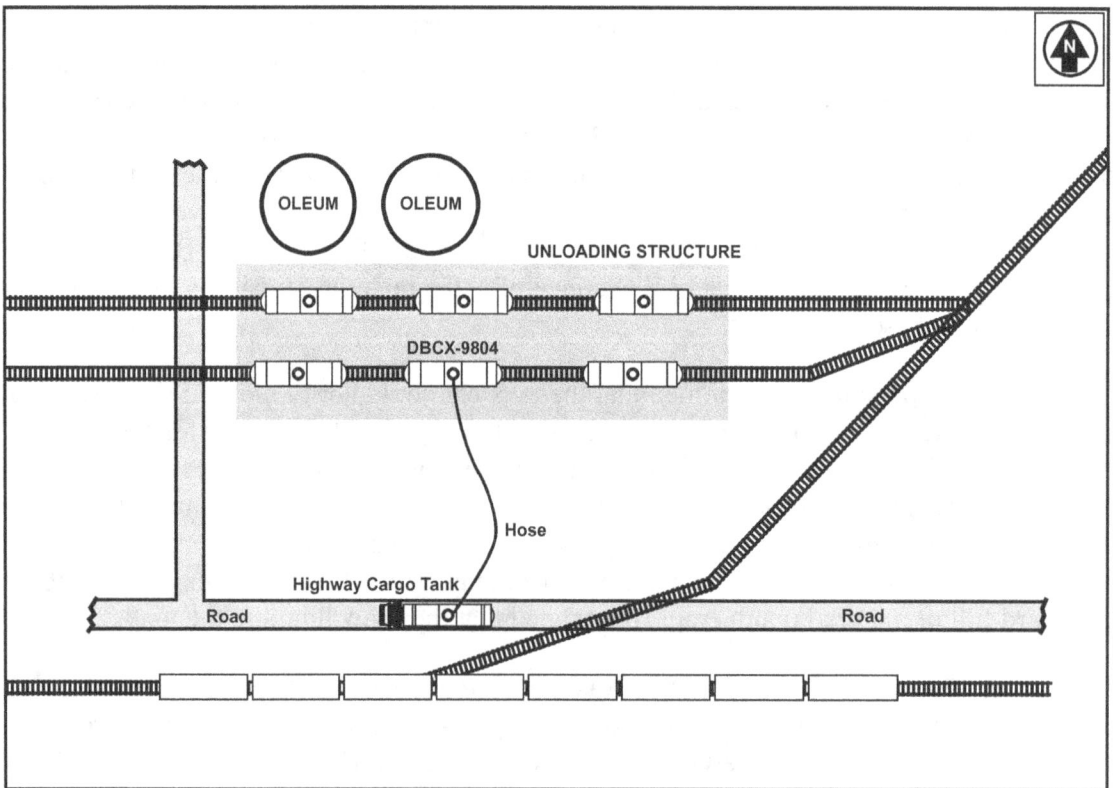

Figure 2. Schematic drawing of Freeport facility transfer area showing the highway cargo tank, railroad tank car DBCX 9804, unloading hose, and oleum tanks.

---

[13] A *vapor line valve* opens to the vapor space within the tank car.

On September 12, about 7:00 a.m., unloaders again applied 60-psig steam to the tank car. The steam-heating continued until 12:00 p.m., when a transfer was attempted. The material did not transfer. All transfer hose connections were left in place. The primary unloader told a BASF operations coordinator that the product was too thick to unload and suggested continuing the application of steam heat until the next morning. The operations coordinator agreed. According to the primary unloader, about 2:00 p.m., a steam trap[14] was added to the tank car's outlet coil, and the steam pressure was reduced to about 20 psig. The heating of the tank car continued uninterrupted for 17 hours, until 7:00 a.m. the following morning. No one monitored the temperature of the material or the pressure within the tank car during the overnight heating, although workers were in the area through the night.

On September 13, about 7:00 a.m., the primary unloader stopped heating the tank car. He was able to transfer product from the tank car to the highway cargo tank at this time. About 7 minutes after the transfer began, the primary unloader saw a splash from the highway cargo tank's manway opening, indicating that the level of the material in the highway cargo tank had reached the transfer hose. The transfer was stopped, and the 3-inch valve on top of the tank car was closed. The BASF operations coordinator climbed on top of the highway cargo tank, looked in its manway opening, and decided that the highway cargo tank was full.

Tank Car Rupture

About 7:45 a.m., the primary unloader and a secondary unloader (who had been present during the unloading) left the area to do other tasks. About 8:40 a.m., the secondary unloader returned to the unloading area. He noticed that the tank car's pressure relief valve had opened and then reset itself. He called the primary unloader and asked what should be done. The primary unloader said the tank car just needed to vent and that it should be left alone.

According to the secondary unloader, the pressure relief valve opened and reset one or two more times. The first time, the valve opened for 3 or 4 minutes and then closed for about 5 minutes. The following time(s), the valve closed for several minutes between openings. Numerous workers noticed an ammonia smell coming from the car.

About 9:00 a.m., the tank car's pressure relief valve began to continuously vent white vapor. (The continuous venting persisted until the tank rupture occurred.) Facility personnel called the Dow Chemicals fire department for assistance.[15] Witnesses reported hearing a high-pitched whistle sound coming from the relief valve around this time. BASF activated a "seek shelter" horn in an adjacent process.

---

[14] A *steam trap* recirculates the steam within the heating coils while allowing the condensate to flow out of the coils. A steam trap increases the efficiency of the steam-heating process.

[15] A Dow Chemicals facility is adjacent to the BASF facility in Freeport. BASF had contracted with the Dow facility for Dow to provide BASF some limited emergency services.

At 9:05 a.m., BASF personnel began to apply water to the tank car to knock down the venting vapors and to cool the tank car. At 9:10 a.m., evacuation horns within the caprolactam II process sounded, and all nonessential employees were ordered to evacuate the area. At 9:15 a.m., Dow Chemicals fire department personnel arrived and began applying water to the tank car using their fire truck. At 9:25 a.m., unmanned fire hoses were positioned to apply water to the tank car while on-scene personnel moved away from the area to discuss further actions. About 9:30 a.m., while water was being applied, the tank car catastrophically ruptured.

Emergency Response

Eleven emergency response agencies responded to the accident, including local fire departments, law enforcement agencies, emergency medical services organizations, a public safety agency, and an emergency management department. After the accident, residents living within 1 mile of the accident site were ordered to shelter in place. About 5 1/2 hours later, the shelter-in-place order was lifted. The area immediately surrounding the accident site (including several facility processes) at the BASF facility remained evacuated, except for essential emergency responders, for about 10 days after the accident due to a continuing leak of the hazardous material oleum from damaged storage tanks and the danger from structural damage to the transfer station.

Injuries

No fatalities were caused by the accident. According to a BASF nurse from the Freeport facility, 28 people reported minor injuries and all these individuals were treated and released from medical supervision.

Damage

Tank car DBCX 9804 split longitudinally near its top. The split ran the length of the tank car. Two circumferential tears on opposite sides of the car ran from the area where the dome housings had been to about halfway down the sides of the tank car. The tank's fracture surfaces were consistent with overstress fracturing. The sides of the tank car had unfolded into a flattened position. (See figure 3.) The pressure relief valve was not found. One dome housing (without the cover) that contained the valves and fittings was found on the adjacent Dow Chemicals facility property about 1/3 mile south of the accident site. The second dome housing (containing the 6-inch access opening) was not found. Both tank car heads were separated from the tank car shell. The B-end head was found about 25 feet west of the flattened tank car, and the A-end head was found about 125 feet to the east.

Figure 3. Postaccident photo showing a Safety Board investigator standing on flattened portion of railroad tank car DBCX 9804.

The transfer station where DBCX 9804 had been, the highway cargo tank that had been receiving the waste, and a second tank car in the vicinity were destroyed. The station structure east and west of the blast site was severely damaged. Two storage tanks containing oleum immediately to the north/northwest of the accident scene were damaged by flying debris, leading to the oleum release.

Railroad Tank Car Information

DBCX 9804 was a U.S. Department of Transportation (DOT) specification 111A100W6 railroad tank car manufactured by Trinity Industries in June 1998. The tank car was manufactured for BASF and was intended to transport acrylic acid. (DBCX 9804 had transported 26 shipments of acrylic acid before it began to be used to store waste. The last shipment of acrylic acid made in DBCX 9804 took place on May 10, 2001.) The tank car's capacity was 23,589 gallons.

DBCX 9804 was a jacketed, stainless steel, general-purpose tank car equipped with exterior heating coils. The tank was manufactured of 7/16-inch ASTM A240-type 316L stainless steel. The tank car had glass wool and ceramic fiber blanket insulation covered by a steel jacket. (See figure 4 to view a comparable tank car.)

Figure 4. Undamaged tank car comparable to DBCX 9804.

According to postaccident estimates calculated by Trinity Industries, the burst pressure of DBCX 9804 was about 595 psig. When the tank car was manufactured, its tank and pressure relief valve were pressure-tested to 100 psig and 75 psig, respectively.

The valves, gauging devices, and tank access points were inside the two dome cover housings on the top of the tank car. (See figure 5.) One housing contained a 20-inch-diameter cover assembly. A 6-inch-diameter access plate was secured to the top of the 20-inch cover assembly. The opening through the 6-inch flange extended through the 20-inch cover assembly, which allowed for visual inspection of the tank interior through the 6-inch access plate opening.

Figure 5. Photo showing interiors of two dome housings of tank car comparable to DBCX 9804. Housing to the right contains 6-inch access opening and 20-inch access opening.

Product transfer took place by way of the valves and fittings located within the second dome cover housing. (See figure 6.) The second dome housing contained the eduction line valve, which was used for loading and unloading the tank car. A 3-inch-diameter eduction line extended from the eduction valve to near the bottom of the tank car. The dome housing also contained the 2-inch-diameter vapor line valve, which could be used to pressurize the tank car with nitrogen.

Figure 6. Housing of second dome on tank car comparable to DBCX 9804.

The pressure relief valve was mounted on top of the tank car, outside of the dome cover housings. The pressure relief valve, a Midland Manufacturing model A-1779-P-MO-CZ, had a set pressure of 75 psig and a flow rating of 1,340 standard feet per minute, air, at 75 psig.

During postaccident examination of the tank car, investigators found no indications of corrosion or unusual wear. Since the tank car's manufacture, only its brakes had undergone repairs. No repairs to the tank or its attachments were recorded.

DBCX 9804 had last been cleaned on June 9, 2001. The cleaning consisted of water-rinsing the tank for 4 1/2 hours and then removing the water by vacuum.

Hazardous Materials Information

The waste material in DBCX 9804 was a combination of cyclohexanone oxime (oxime),[16] water, and cyclohexanone. The majority of the waste material (94 percent) was

---

[16] BASF typically referred to this material as "oxime," so from this point onward, this report will generally use this term to refer to cyclohexanone oxime.

oxime, which is a white, solid material under normal atmospheric conditions at temperatures below 194° F. When solid, oxime is not classified as a hazardous material by the DOT but, when liquid, oxime is flammable. The material safety data sheet for oxime produced by BASF in 1994 describes the material as stable under normal temperatures and pressures but states that temperatures above 212° F should be avoided.

About 4 percent of the waste material was water.

About 2 percent of the waste material was cyclohexanone, which is a flammable liquid with a flashpoint of 111° F. Its flammable limits are 1.1 percent to 8.1 percent in air. The BASF material safety data sheet for cyclohexanone describes the material as stable.

Oxime and cyclohexanone are maintained between 180° F and 204° F within the caprolactam production process. The process operates at atmospheric pressure.

BASF Oxime Testing

Postaccident

BASF's postaccident analysis of the material from the tank car showed that its chemical constituents were consistent with the oxime waste material from the Freeport process upsets. BASF found no additional constituents in the material.

Because the events preceding the accident (the heating of the tank car and the continuous venting of the tank car's pressure relief valve) suggested that overpressurization from a chemical reaction might have brought about the accident, BASF conducted postaccident Automatic Pressure Tracking Adiabatic Calorimeter tests on residual oxime material from the accident scene to determine why the tank car ruptured.[17] For comparison purposes, BASF ran the same tests on pure oxime from several different sources. The tests measured changes in the oxime's temperature and pressure as it was heated.

The BASF test results showed that untempered oxime had a significantly higher onset temperature[18] and a longer induction time[19] than tempered[20] oxime. In effect, BASF found that tempering oxime reduced its stability.

---

[17] A Safety Board chemist reviewed the testing methodology and parameters used by BASF during its postaccident testing and found them reasonable and appropriate.

[18] The *onset temperature* is the lowest temperature at which exothermic (heat-releasing) behavior is exhibited.

[19] The *induction time* is the time required, at a given temperature above the onset temperature, for the decomposition reaction to accelerate to an uncontrolled runaway condition.

[20] Oxime is tempered by putting it through cycles of heating and cooling.

BASF tested one sample of untempered oxime obtained from a common chemical supplier. Water was added to bring the percentage of water in the sample to 4.5 percent. When the sample was heated,[21] exothermic reactions were first observed at 338° F, about 590 minutes into the experiment. A second spike in exothermic reactions was recorded after 895 minutes.

BASF also tested a sample of tempered oxime material (containing about 2.8 percent water) recovered from the highway cargo tank at the accident scene. When heated, the sample first showed exothermic reaction at 282° F, about 503 minutes into the experiment. More exothermic activity was recorded after about 665 minutes.

To confirm the lower onset temperature for tempered oxime, BASF conducted a further test. BASF took a sample of oxime[22] from the caprolactam process at the Freeport facility, tempered the material at 230° F for 840 minutes, and then allowed it to cool. When this sample was heated as the previous two samples had been, BASF recorded two episodes of exothermic activity–the first at 258° F and the second starting at 381° F.

On the basis of its testing, BASF determined that the following reactions occurred when oxime was heated:

1) Cyclohexanone Oxime + Heat + H_2O ------$\rightarrow$ Cyclohexanone + NH_2OH
   (hydroxylamine).
Then,
2) $4NH_2OH$ + Heat ------$\rightarrow$ N_2O (nitrous oxide gas) + $2NH_3$ (ammonia gas) + $3H_2O$.

BASF further determined that the second reaction is essentially irreversible because two of the products of the reaction are gases and a state of equilibrium is never attained. BASF found that once the first reaction started, both reactions would continue until the reactants were consumed.

BASF also conducted five analyses of the gases that evolved during the heating of the oxime. The results of all five tests showed the presence of ammonia and nitrous oxide.

Preaccident

From 1989 to 1991, the BASF parent plant in Ludwigshafen, Germany, conducted analytical tests of oxime. The data generated by the tests included information concerning onset temperature and induction time. The data showed that the onset temperatures for oxime ranged from 248° F to 338° F. The induction times for oxime ranged from 480 minutes at 311° F to 1,320 minutes at 275° F. Data from the tests were discussed at a

---

[21] In the postaccident testing BASF conducted to determine when exothermic reactions would occur for the various oxime samples, the samples were initially heated to 212° F, the maximum safe temperature for oxime, before the test heating process was begun.

[22] The sample was about 4.5 percent water.

1991 BASF caprolactam technology exchange meeting in Belgium, which personnel from the Freeport facility attended. Some of the data appeared in an appendix of the *BASF Technical Exchange Packet for 1991*, which was distributed at the meeting. The stated goal of the appendix was to "prevent an out of control reaction resulting in emissions of anone, hydroxylamine and/or oxime." The appendix stated, "To avoid uncontrolled rearrangement, temperature in oxime tank should not exceed 194° F. Proper attention is given to the explosion limits of oxime in air."

BASF Information

Company and Facility Information

The BASF Group comprises a parent plant in Ludwigshafen and 164 wholly owned subsidiaries. The BASF Group has production facilities in 38 countries and manufactures a wide range of products, including chemicals, polymers, automotive and industrial coatings, colorants, vitamins, nylon fibers, and agricultural products. The BASF Corporation is the North American affiliate of the BASF Group.

More than 700 employees and 350 contractors work at the 406-acre Freeport BASF facility. The facility produces 23 different products in 16 processes. Raw materials arrive at the facility by rail, highway, barge, and pipeline. Intermediate chemicals produced at the facility are shipped out by rail, highway, and barge.

Procedures for Heating and Unloading Hazardous Materials at Freeport

According to BASF officials, the Freeport facility had no written procedures for heating and unloading the specific waste material contained in tank car DBCX 9804. The oxime mixture is an intermediate material in the production process for caprolactam, and it is not typically removed from the process or shipped for disposal from the Freeport facility. Consequently, the Freeport unloaders did not routinely handle this material. According to BASF, the last shipment of oxime from Freeport had taken place more than 10 years before the accident, and the material had never before been shipped from the facility in a railroad tank car.

The BASF Freeport facility had general operating procedures for transferring material from a highway cargo tank to a railroad tank car and for using pressure to transfer material from a tank car to a plant receiving tank. The facility also provided product-specific unloading and heating procedures for the two materials that are routinely steam-heated in tank cars at the Freeport facility—caprolactam and formaldehyde.

Of the written procedures provided by the BASF Freeport facility, only the general operating procedure for transferring material from a tank car to a receiving tank required that the material be sampled and tested by lab personnel before the transfer could begin. Only the product-specific procedures for steam-heating formaldehyde required that the pressure within the tank car be monitored during the steam-heating process. Those procedures specified that the pressure within the tank car be checked every 30 minutes. Neither the procedures for steam-heating formaldehyde nor those for steam-heating caprolactam specified the pressure or temperature of the steam to be applied. The Freeport facility did not provide the unloaders with additional or specific procedures for unloading DBCX 9804.

BASF Postaccident Actions

Since the accident, the BASF Freeport facility has made several changes to its facility procedures. It has developed and implemented written procedures specifically addressing the safe handling of oxime waste material. The procedures prohibit the transfer of oxime waste to railroad tank cars and the heating of closed containers of oxime. The Freeport facility now requires that hot water, rather than steam, be used to heat oxime waste. Additionally, BASF changed its material safety data sheet for oxime to specify that the material must not be heated above 212° F.

Analysis

This analysis has three main sections. First, it identifies factors that can be readily excluded as causal or contributory to the accident. Next, the analysis addresses the accident and its possible causes. The remainder of the analysis discusses the following safety issue arising from this investigation:

- Adequacy of procedures for heating hazardous materials cargoes in railroad tank cars before transfer.

Exclusions

Before the Freeport accident, railroad tank car DBCX 9804 had not been in any accidents, and the past repairs to this tank car involved only its brakes. Postaccident examination of the tank car showed no evidence of corrosion or other signs of deterioration. Postaccident evaluation of the tank car's fracture surfaces indicated that they were consistent with overstress failure. Therefore, the Safety Board concludes that railroad tank car DBCX 9804 had no structural or material defects that caused or contributed to the rupture of the tank.

The Accident

Rupture of DBCX 9804

Although the pressure relief valve for DBCX 9804 was not recovered after the accident, the events described by the witnesses at the scene on September 13, 2002, indicate that the pressure relief valve was functioning and venting vapors up to the time the tank car ruptured. Witnesses recalled that the pressure relief valve opened and reset itself two or three times between about 8:40 a.m. and 9:00 a.m., began to constantly vent about 9:00 a.m., and continued to constantly vent until the rupture occurred about 9:30 a.m. Witnesses also reported hearing a high-pitched whistle around 9:00 a.m. and seeing a stream of white vapor come from the pressure relief valve as it constantly vented. During postaccident inspection of the tank car, investigators found a longitudinal tear along the top of the tank car and circumferential tears near the dome housings.

The initial cycling of the pressure relief valve's opening and closing, followed by its continuous venting up to the time the tank car ruptured, indicates that the gas within the tank car was being produced more rapidly than it could be vented. Further, this activity of the pressure relief valve and the postaccident condition of the tank car—including the flattened and torn tank shell, the damage to the tank heads, and the projection of one dome housing a substantial distance from the tank car—strongly indicate that an overpressure condition developed due to the rapid generation of gas or vapor within the tank from a chemical reaction involving the hazardous waste in the tank car.

The DOT hazardous materials regulations (49 CFR, Subchapter C) require that tank car tanks have a pressure relief device with a sufficient flow capacity to prevent pressure in the tank from exceeding a maximum pressure under specified fire conditions. The regulations do not specifically require that pressure relief devices have the flow capacity to relieve pressure generated from a chemical reaction within the tank. Rather, the regulations include requirements that are intended to prevent chemical reactions from occurring. For example, all components of a tank car must be chemically compatible with the chemical cargoes authorized for carriage. Hazardous materials that self-react or are highly reactive must have chemical stabilizers added to prevent these adverse reactions from occurring during transportation. The Safety Board considers that this approach is reasonable, and it has proven to be effective. Therefore, the Safety Board concludes that the pressure relief valve on tank car DBCX 9804 functioned as designed and vented vapors continuously once it activated, but the rapid generation of gas within the tank exceeded the valve's capacity to relieve the tank's internal pressure sufficiently to prevent the catastrophic rupture of the tank car.

Overpressure Condition

The Safety Board considered several possible causes for the overpressurization that led to the tank car's rupture, including 1) instability of the waste material, 2) introduction of chemically incompatible material to the waste material within the tank, and 3) an adverse reaction resulting from the heating of the waste in the tank car.

Instability of the Waste Material. BASF material safety data sheets for oxime and cyclohexanone state that both materials are stable. In addition, the waste had been in the tank car for more than a year—experiencing significant changes in ambient temperatures and humidity—and had arrived at the MFR plant in Hannibal, Missouri, without any reported abnormalities, such as elevated tank pressure or material temperature. Further, the tank car was shipped back from Hannibal to Freeport without incident, and it remained in Freeport without unusual developments for 2 months before the accident. Thus, instability of the material did not appear to be the cause of the overpressurization.

Introduction of Incompatible Material. Another possible cause of the overpressurization in the tank was the introduction of a chemically incompatible material, either intentionally or unintentionally, which caused the reaction in the tank that led to the overpressurization. Given that the tank car arrived in Freeport without the security seals that MFR said it had applied in Hannibal, the introduction of an incompatible material during the transport period may have been possible. However, the routing of the tank car from Hannibal to Freeport was traced, and no unusual delays were found. Further, no indications of product tampering were found.

The waste mixture was not reactive with steels or metals, and postaccident analysis of the material from the tank car showed that the chemical constituents were consistent with the waste material from the process upsets in Freeport. Therefore, it appears that the overpressurization was not caused by the introduction of incompatible chemical contaminants.

Heating the Tank Car. BASF's postaccident tests of oxime materials showed that heat-tempered oxime like the material in DBCX 9804 has a lower onset temperature than untempered oxime. Consequently, less heat would have been needed to initiate an exothermic (heat-releasing) reaction of the tempered material in DBCX 9804 than would have been necessary for untempered material. Specifically, the tests showed that untempered oxime required heating to about 338° F to show the first signs of exothermic reaction. In contrast, the tempered oxime from the accident scene, as well as another sample of tempered oxime, required heating only to about 282° F to show the first signs of exothermic reaction. Thermodynamic tables show that, for the heating conducted at the BASF facility, 60-psig steam would have had a minimum temperature of 308° F and 20-psig steam a minimum temperature of 260° F. (BASF applied 60-psig steam to the tank for about 1 hour on September 11 and about 7 hours on September 12. BASF applied 20-psig steam to the tank for about 17 hours from September 12 through the early morning of September 13.)

According to the results of BASF postaccident testing, given the amount of waste material in DBCX 9804 at the time of the accident and the steam pressures used by BASF to heat the material in Freeport, 23 hours of heating would have been required to raise the temperature of the material in the tank car to 284° F, surpassing the estimated onset temperature of 282° F. In fact, BASF heated the material for about 24 hours on September 12 and 13. The use of the steam trap during the heating would have increased the efficiency of the steam-heating process.

Further testing showed that when heated, tempered oxime broke down into cyclohexanone and hydroxylamine. The heat released from this breakdown reaction was sufficient to initiate a second exothermic reaction, the decomposition of hydroxylamine into nitrous oxide and ammonia gases. The generation of ammonia from such a reaction is consistent with the witness reports of ammonia odors as DBCX 9804 vented before the accident.

Therefore, the steam heat that BASF applied to DBCX 9804 in the hours preceding the accident at Freeport was sufficient to initiate two exothermic reactions of the waste material. Both reactions released heat in an insulated tank, the insulation of which likely prevented the heat's dissipation. This resulted in the generation and retention of heat at a rate that led to a runaway reaction and build-up of pressure within the tank. Given the runaway nature of the reaction, the capacity of the pressure relief valve to vent the gases and relieve the pressure in the tank was exceeded, so the pressure increased until the tank ruptured. The Safety Board concludes that the overpressurization and rupture of the tank car was caused by a runaway exothermic decomposition reaction that resulted when the oxime waste material the tank car contained was steam-heated to an excessive temperature for an extended period of time.

Procedures Used in Steam-Heating DBCX 9804

In the months preceding the accident, both MFR and BASF had steam-heated tank car DBCX 9804 to liquefy the hardened waste material within it so that the waste could be transferred from the tank car. However, MFR and BASF followed different procedures while performing these operations.

MFR Procedures

Different waste materials may have widely dissimilar chemical make-ups and properties. As might be expected of a company that unloads many different waste materials of diverse compositions, MFR was not familiar with the specific chemical properties of the oxime waste material in DBCX 9804. (MFR had never previously unloaded this particular waste material from a tank car.) MFR knew only the waste's heating value and that it did not contain metals.

Because MFR did not have detailed information about the chemical properties of the waste, it was concerned about the potential for an increase in vapor pressure that could result from heating the tank car. Consequently, MFR took precautions to monitor the interior conditions of the tank car during heating. MFR used a gauging device to keep track of the temperature and pressure within the tank car, employed a safety mechanism to automatically stop the heating of the tank car if the pressure reached a preset limit (15 psig), periodically conducted visual inspections of the material, and halted the application of heat when no employees were present (at night) to monitor the process. MFR records show that the pressure within the tank car never exceeded 3.5 psig while the tank car was being heated at the Hannibal facility. However, had a significant increase in the internal tank pressure developed, the MFR procedures for monitoring the heating process probably would have alerted MFR employees before the pressure reached critical levels. In such a situation, MFR would most likely have had sufficient time to take appropriate measures to relieve the pressure within the tank car (such as by opening the 6-inch access plate) before the pressure could reach a level that would rupture the tank.

It should be noted that although MFR used appropriate safety procedures when heating DBCX 9804, MFR's lack of procedures for ensuring that the tank car had been emptied of all waste material after the unloading process at Hannibal was terminated allowed the tank car to be sent back to BASF containing about 8,000 to 10,000 gallons of waste. Had MFR taken steps to ensure that the tank car was truly empty before sending it back to BASF (such as by visually inspecting the interior of the tank through the 6-inch access opening), MFR would have learned that the tank car still contained a significant amount of material. Then, MFR could have unloaded the remaining waste and sent DBCX 9804 back to Freeport empty, in which case the Freeport facility would not have had to deal with the disposal of the material.

BASF Procedures

Like the unloading personnel at MFR, the unloaders in Freeport did not routinely heat oxime waste material for transfer. According to BASF, the Freeport facility had last shipped this material more than 10 years before the accident. And, although the oxime material was an intermediate product from the Freeport facility, it had never before been heated and transferred from a railroad tank car. Consequently, the BASF Freeport facility had no specific written procedures for heating the waste mixture before unloading it from the tank car. Perhaps the Freeport facility could not have been expected to have predetermined procedures for heating this particular hazardous waste mixture, but the facility's lack of experience with the material should have indicated that special care should be taken during the heating process.

Instead of adopting reasonable safety measures, the BASF Freeport facility treated the material as if it posed little or no risk. Specifically, the unloaders in Freeport did not use a gauge or other device to monitor the temperature and pressure within the tank; they did not apply any safety mechanism that would stop the heating if the pressure inside the tank reached a preset limit; and they did not periodically evaluate the tank's interior conditions.

The unloaders at the Freeport facility did not use any monitoring tools to help them assess the internal conditions of the tank during the heating process. Because the facility did not require its unloading personnel to use a pressure gauge to monitor the pressure in the tank car during the heating process, the first indication the unloaders had of the rising pressure in the tank car was the activation of the pressure relief valve when the pressure exceeded 75 psig. This happened about 45 minutes before the tank car ruptured. By this time, it was likely too late to stop the runaway chemical reaction.

Nor did the Freeport unloaders use a safety mechanism to automatically stop the heating if the pressure reached a preset limit. Had they used an automatic shutdown mechanism, the tank car heating would likely have been terminated much sooner, when the tank pressure rose to the established safe limit.

Further, if the Freeport facility workers had monitored and evaluated the conditions within the tank car (specifically, the temperature of the waste material and the internal pressure of the tank car) by checking them periodically, they would have found that the material was sufficiently liquid for successful transfer around 9:00 p.m. on September 12—about 12 1/2 hours before the accident occurred. They also would have had time to relieve the pressure within the tank car before it reached critical levels, by taking such action as opening the 6-inch access plate.

Therefore, the Safety Board concludes that lack of experience in transferring the oxime material should at least have led the BASF Freeport facility, like MFR, to take precautions during the heating process.

Although the workers immediately involved in the transfer operation may not have been aware of the specific properties of the oxime waste they were unloading, BASF had such information. The 1994 BASF material safety data sheet for oxime specifically stated that temperatures above 212° F should be avoided for this material. Additionally, during a 1991 BASF meeting in Belgium, the variations in onset temperatures for oxime were discussed. Guidance distributed to those attending the meeting (including representatives from the Freeport facility) advised that to avoid uncontrolled runaway reactions of tempered oxime, the temperature within the tank should not exceed 194° F. The Safety Board concludes that, before the accident, BASF had information concerning the dangers associated with excessive heating of oxime that should have alerted the Freeport facility to the need for developing and implementing safe procedures for heating and unloading this material from railroad tank cars.

The BASF Freeport facility has changed its facility procedures to address problems identified during the investigation of this accident. It has developed and implemented written procedures designed to ensure the safe handling of oxime waste material. The procedures prohibit the transfer of oxime waste to railroad tank cars, as well as the heating of closed containers of oxime. Additionally, the Freeport facility now requires that hot water be used, rather than steam, to heat oxime waste. BASF revised its material safety data sheet for oxime to stipulate that oxime must not be heated above 212° F. The Safety Board considers that these actions will significantly reduce the likelihood of incidence of an accident similar to the one discussed in this report. Consequently, the Safety Board makes no recommendations to the BASF Freeport facility concerning its oxime handling procedures.

Heating Hazardous Materials in Railroad Tank Cars

The Safety Board addressed issues involving the heating of hazardous materials cargoes prior to unloading in its investigation of the rupture of a railroad tank car that took place in Clymers, Indiana, on February 18, 1999.[23] The Safety Board determined that the tank car had overpressurized and catastrophically ruptured after a waste material had been heated in it for more than 28 hours over several days. As in the Freeport accident, the internal tank conditions were not monitored during the heating process. In the Clymers investigation, the Safety Board determined that the failures in the procedures were attributable to the companies involved with the shipment and heating. Consequently, the Safety Board issued the following safety recommendations to the specific companies involved with the generation and disposal of the waste material in the Clymers accident:

I-01-2 through -5

Collaborate with applicable producers, shippers, consignees, and end-users in the development and implementation of specific and written procedures for the loading or offloading of any chemical or waste material from a railroad tank car, highway cargo tank, or other bulk transportation vessel when the chemical or waste material exhibits properties that require special handling or processing during the loading or offloading operation.

---

[23] National Transportation Safety Board, *Rupture of a Railroad Tank Car Containing Hazardous Waste Near Clymers, Indiana, February 18, 1999*, Hazardous Materials Accident Report NTSB/HZM-01/01 (Washington, DC: NTSB, 2001).

Two of the four companies receiving the safety recommendations did not respond. Safety Recommendations I-01-2 and -3 were therefore classified as "Closed—Unacceptable Action—No Response Received" on July 31, 2003. The third company responded that it considered the recommendation unwarranted. Consequently, Safety Recommendation I-01-4 was classified as "Closed—Unacceptable Action" on September 12, 2001. The fourth company ultimately determined not to implement Safety Recommendation I-01-5, which was classified as "Closed—Unacceptable Action" on September 16, 2004. Because of the lack of action by the companies involved in the Clymers accident, as well as the circumstances of the Freeport accident, the Safety Board is concerned that the heating of hazardous materials cargoes in railroad tank cars remains a significant safety problem.

The DOT hazardous materials regulations[24] do not specifically address the heating of cargo in tank cars. On October 30, 2003, the Research and Special Programs Administration (RSPA) published a final rule under docket HM-223 that was intended to clarify the applicability of the hazardous materials regulations to the loading, unloading, and storage of hazardous materials during transportation. Under this final rule, with respect to tank cars and other bulk containers, RSPA has interpreted "unloading incidental to movement," which is subject to the DOT hazardous materials regulations, to take place only when a hazardous material is emptied from its bulk packaging after the material has been delivered to the consignee and prior to the delivering carrier's departure from the consignee's facility or premises. Because virtually all tank cars are unloaded by the consignee after the delivering rail carrier has departed, this rule in effect means that RSPA no longer considers the unloading of a tank car to be a RSPA-regulated transportation function.

The final rule under HM-223 was to have become effective on October 1, 2004. In a *Federal Register* notice published on May 28, 2004, RSPA stated it was delaying the effective date to January 1, 2005. In the notice, RSPA stated that it had received 14 appeals to the final rule. According to RSPA, the appellants raised a number of issues concerning the consistency of the final rule with Federal hazardous materials transportation law, State and local regulation of hazardous materials facilities, the relationship of the DOT hazardous materials regulations to Occupational Safety and Health Administration (OSHA) and Environmental Protection Agency (EPA) regulations, the definition of "unloading incidental to movement," and other aspects of the final rule. Also, 11 industry associations filed a petition on December 29, 2003, in the U.S. Court of Appeals for the District of Columbia to reverse the HM-223 final rule.

The Safety Board initially summarized its concerns in an October 29, 2001, response to the Notice of Proposed Rulemaking for HM-223 by stating that the proposed rule may result in the elimination of effective Federal oversight of hazardous materials loading/unloading operations for bulk transportation containers. The Safety Board also emphasized in its response that the Board has historically and consistently considered loading and unloading operations, particularly of bulk containers such as tank cars, to be

---

[24] The DOT's Research and Special Programs Administration promulgates these regulations.

transportation-related functions. The Board also expressed its belief that the DOT has both the statutory mandate and the authority to regulate loading and unloading operations, and that the DOT should strengthen its oversight of these operations rather than ignore these issues.

The Safety Board's concerns were reinforced as a result of its investigation of the July 14, 2001, accident at a chemical plant in Riverview, Michigan.[25] In this accident, methyl mercaptan, a poisonous and flammable gas, was released and ignited during the offloading of a railroad tank car. Three plant employees were killed and about 2,000 residents were evacuated from their homes for 10 hours. Although the unloading operations involving the accident tank car were subject to OSHA's process safety management and the EPA's risk management programs, the Safety Board concluded that effective oversight of hazardous materials loading and unloading operations from tank cars and other bulk containers was not provided by the Federal Railroad Administration, the EPA, or OSHA. As a result, the Safety Board issued the following safety recommendations to the DOT:

I-02-1

Develop, with the assistance of the EPA and OSHA, safety requirements that apply to the loading and unloading of railroad tank cars, highway cargo tanks, and other bulk containers that address the inspection and maintenance of cargo transfer equipment, emergency shutdown procedures, and personal protection requirements.

I-02-2

Implement, after the adoption of safety requirements developed in response to Safety Recommendation I-02-1, an oversight program to ensure compliance with these requirements.

The Safety Board also issued Safety Recommendations I-02-3 and -4 to OSHA and the EPA, respectively, to recommend that they assist the DOT in the development of the safety requirements referred to in Safety Recommendations I-02-1 and -2.

In a November 25, 2002, response to Safety Recommendations I-02-1 and -2, RSPA noted that it had worked closely with OSHA and the EPA on various hazardous materials issues and that it was proceeding with the HM-223 rulemaking. In a February 25, 2003, reply to RSPA's response, the Safety Board again expressed its concern that the rulemaking could have an adverse effect on the safety of loading and unloading operations, and the Board classified both recommendations "Open—Unacceptable Response." OSHA and the EPA responded to Safety Recommendations I-02-3 and -4 in August and October 2002, respectively. The two agencies pledged to work with the DOT to develop the needed safety requirements. As a result, the Safety Board classified Safety Recommendations I-02-3 and -4 "Open—Acceptable Response."

---

[25] National Transportation Safety Board, *Hazardous Materials Release from Railroad Tank Car With Subsequent Fire at Riverview, Michigan, July 14, 2001*, Hazardous Materials Accident Report NTSB/HZM-02/01 (Washington, DC: NTSB, 2002).

The Safety Board's investigation of the Riverview accident identified a lack of effective oversight of hazardous materials loading and unloading operations with respect to inspection and maintenance of equipment, emergency shutdown procedures, and personal protection. Further, the accidents in Clymers and Freeport identify a lack of Federal oversight when hazardous materials in railroad tank cars are heated prior to unloading the materials. In the absence of any apparent Federal regulations addressing this issue, and given the uncertainty over the eventual outcome of the appeals to HM-223, the Safety Board concludes that Federal oversight of operations for heating hazardous materials in railroad tank cars prior to unloading is inadequate. The Safety Board believes that RSPA should, in cooperation with OSHA and the EPA, develop regulations that require safe operating procedures to be established before hazardous materials are heated in a railroad tank car for unloading; at a minimum, the procedures should include the monitoring of internal tank pressure and cargo temperature.

Conclusions

Findings

1. Railroad tank car DBCX 9804 had no structural or material defects that caused or contributed to the rupture of the tank.

2. The pressure relief valve on tank car DBCX 9804 functioned as designed and vented vapors continuously once it activated, but the rapid generation of gas within the tank exceeded the valve's capacity to relieve the tank's internal pressure sufficiently to prevent the catastrophic rupture of the tank car.

3. The overpressurization and rupture of the tank car was caused by a runaway exothermic decomposition reaction that resulted when the cyclohexanone oxime waste material the tank car contained was steam-heated to an excessive temperature for an extended period of time.

4. Lack of experience in transferring the cyclohexanone oxime material should at least have led the BASF Freeport facility, like Missouri Fuel Recycling Environmental Services, to take precautions during the heating process.

5. Before the accident, the BASF Corporation had information concerning the dangers associated with excessive heating of cyclohexanone oxime that should have alerted the Freeport facility to the need for developing and implementing safe procedures for heating and unloading this material from railroad tank cars.

6. Federal oversight of operations for heating hazardous materials in railroad tank cars prior to unloading is inadequate.

Probable Cause

The National Transportation Safety Board determines that the probable cause of the rupture of railroad tank car DBCX 9804 was overpressurization resulting from a runaway exothermic decomposition reaction initiated by excessive heating of a hazardous waste material. Contributing to the accident was the BASF Corporation's failure to monitor the temperature and pressure inside the tank car during the heating of the hazardous waste.

Recommendations

As a result of its investigation of the September 13, 2002, hazardous materials accident at Freeport, Texas, the National Transportation Safety Board makes the following safety recommendations:

To the Research and Special Programs Administration:

In cooperation with the Occupational Safety and Health Administration and the Environmental Protection Agency, develop regulations that require safe operating procedures to be established before hazardous materials are heated in a railroad tank car for unloading; at a minimum, the procedures should include the monitoring of internal tank pressure and cargo temperature. (R-04-10)

To the Occupational Safety and Health Administration:

In cooperation with the Research and Special Programs Administration and the Environmental Protection Agency, develop regulations that require safe operating procedures to be established before hazardous materials are heated in a railroad tank car for unloading; at a minimum, the procedures should include the monitoring of internal tank pressure and cargo temperature. (R-04-11)

To the Environmental Protection Agency:

In cooperation with the Research and Special Programs Administration and the Occupational Safety and Health Administration, develop regulations that require safe operating procedures to be established before hazardous materials are heated in a railroad tank car for unloading; at a minimum, the procedures should include the monitoring of internal tank pressure and cargo temperature. (R-04-12)

BY THE NATIONAL TRANSPORTATION SAFETY BOARD

MARK V. ROSENKER
Vice Chairman

CAROL J. CARMODY
Member

RICHARD F. HEALING
Member

DEBORAH A. P. HERSMAN
Member

Adopted: December 1, 2004

Chairman Ellen Engleman Conners did not participate in the adoption of this report.

this page intentionally left blank

Appendix A

Investigation

The National Transportation Safety Board was notified of the accident about 10:30 a.m. eastern daylight time on September 13, 2002. Two hazardous materials investigators were dispatched from Washington, D.C., to Freeport, Texas. No Board Member went with the team.

A single investigative group was established, comprising the following parties: the BASF Corporation; Missouri Fuel Recycling Environmental Services; Trinity Industries, Inc.; the Midland Manufacturing Corporation; the Federal Railroad Administration; and the Railroad Commission of Texas.

The Safety Board did not conduct a public hearing concerning this investigation.

Appendix B

Summary of Significant Events Preceding the Freeport Accident

Date	Event
June 1, 2001	Caprolactam II process upset in Freeport, Texas, BASF facility produces about 91,000 pounds of cyclohexanone oxime mixture waste material
June 5, 2001	BASF transfers the June 1, 2001, process upset waste to four highway cargo tanks at Freeport
June 12, 2001	After heating the waste to liquefy it, BASF transfers the contents of the four highway cargo tanks to railroad tank car DBCX 9804 at Freeport
Between June 12, 2001, and April 26, 2002	BASF applies steam heat to the tank car to liquefy the waste mixture so water can be removed and disposed of on-site
April 2, 2002	BASF loads about 91,000 pounds of cyclohexanone oxime waste from a second caprolactam II process upset at Freeport into four highway cargo tanks (Waste material is chemically indistinguishable from material from June 1, 2001, process upset)
April 26, April 29, and May 10, 2002	BASF transfers the second batch of material from the four highway cargo tanks to DBCX 9804, which still contains the material from the June 1, 2001, caprolactam II process upset
June 21, 2002	BASF sends tank car DBCX 9804, containing the material from both process upsets, to the Missouri Fuel Recycling Environmental Services (MFR) facility in Hannibal, Missouri, so the hazardous waste material can be incinerated
July 7, 2002	DBCX 9804 arrives at the MFR facility in Hannibal
July 9, 2002	MFR steam-heats DBCX 9804 from 9:00 a.m. until 9:30 p.m.
July 10, 2002	MFR steam-heats DBCX 9804 from 5:00 a.m. until 11:00 p.m.
July 11, 2002	MFR unloads waste material from DBCX 9804; MFR mistakenly believes DBCX 9804 is empty
July 12, 2002	MFR sends the partially emptied DBCX 9804 back to the BASF Freeport facility
July 22, 2002	Tank car DBCX 9804 arrives back in Freeport; tank car is judged to be about 1/3 full of waste material
September 11, 2002	BASF begins heating DBCX 9804 (for about 1 hour) so that the material left in the tank can be transferred to highway cargo tanks

Date	Event
September 12, 2002	7:00 a.m. BASF heats DBCX 9804 for about 5 hours 12:00 p.m. BASF attempts to transfer the material from the tank car but is unsuccessful because the material is not sufficiently liquid 2:00 p.m. BASF continues to heat the tank car; heating continues till next morning
September 13, 2002	7:00 a.m. Heating of DBCX 9804 is stopped; transfer of material to highway cargo tanks is begun 7:45 a.m. Transfer of material to one highway cargo tank is completed 8:40 a.m. DBCX 9804 is seen to be venting 9:00 a.m. DBCX 9804 is seen to be continuously venting; fire emergency personnel are summoned from Dow Chemical; the "seek shelter" horn is activated 9:05 a.m. Water is applied to DBCX 9804 to knock down vapors and cool the tank car 9:10 a.m. Area around DBCX 9804 is evacuated 9:15-9:30 a.m. Water is applied to DBCX 9804 9:30 a.m. DBCX 9804 catastrophically ruptures

this page intentionally left blank